W9-AVC-266

FIRE SAFETY

by Emma Bassier

Cody Koala
An Imprint of Pop!
popbooksonline.com

abdobooks.com
Published by Pop!, a division of ABDO, PO Box 398166, Minneapolis, Minnesota 55439.

Printed in the United States of America, North Mankato, Minnesota

052020
092020
THIS BOOK CONTAINS RECYCLED MATERIALS

Cover Photo: Shutterstock Images
Interior Photos: Shutterstock Images, 1, 5 (top), 8–9, 11 (bottom left), 11 (bottom right); iStockphoto, 5 (bottom left), 5 (bottom right), 7, 11 (top), 12, 15, 17, 21; ElenaBs/Alamy, 19

Editor: Connor Stratton
Series Designer: Christine Ha

Library of Congress Control Number: 2019954992

Publisher's Cataloging-in-Publication Data
Names: Bassier, Emma, author.
Title: Fire safety / by Emma Bassier
Description: Minneapolis, Minnesota : POP!, 2021 | Series: Safety for kids | Includes online resources and index
Identifiers: ISBN 9781532167539 (lib. bdg.) | ISBN 9781532168635 (ebook)
Subjects: LCSH: Fire safety--Juvenile literature. | Fire prevention--Juvenile literature. | Fires--Safety measures--Juvenile literature. | Safety education--Juvenile literature. | Accidents--Prevention--Juvenile literature.
Classification: DDC 628.922--dc23

Hello! My name is

Cody Koala

Pop open this book and you'll find QR codes like this one, loaded with information, so you can learn even more!

Scan this code* and others like it while you read, or visit the website below to make this book pop.
popbooksonline.com/fire-safety

*Scanning QR codes requires a web-enabled smart device with a QR code reader app and a camera.

Table of Contents

Chapter 1
Careful Cooking. 4

Chapter 2
Safety at Home 6

Chapter 3
Being Prepared 10

Chapter 4
In an Emergency 16

Making Connections 22
Glossary. 23
Index 24
Online Resources 24

Chapter 1

Careful Cooking

Juan and his family cook dinner together. Flames heat a pot on the stove. The family watches it carefully. When using fire, being careful helps people stay safe.

Watch a video here!

Chapter 2

Safety at Home

Matches, lighters, and candles are in many homes. These items make or use fire. Never play with them. Don't leave them lying around. Ask an adult to help.

Learn more here!

If something catches on fire, call for an adult to help. The adult could use

a **fire extinguisher**. This tool sprays foam. The foam helps put out fires.

Chapter 3

Being Prepared

Even when people are careful, **accidents** can happen. And fires can spread fast after starting. It is important to be prepared.

A fire can spread through a room in less than two minutes.

Complete an activity here!

TEST
DO NOT PAINT

Smoke alarms are devices that **detect** smoke. When there is smoke, the alarms beep loudly. People can put them in bedrooms and around the home. Smoke alarms can quickly warn people if a fire starts.

Families should make an **escape** plan in case of a fire. Talk about how to get out of the house. Choose a place outside where everyone will meet. Then practice. Walk or crawl the path you would take to get out.

Touch doors before you open them. If any part of a door is warm, keep the door shut. Warmth means that flames are right behind the door. When you find an exit, crawl to stay under smoke. Smoke is dangerous to breathe.

What to Do If Clothes Catch Fire

STOP

DROP

ROLL

Find the meeting place when you get outside. Yell for help if people are nearby. Call 9-1-1. Wait for firefighters to arrive. Never go back inside a building that is on fire.

In 2018, the United States had 332,400 firefighters.

Making Connections

Text-to-Self

What is one way you can follow fire safety rules at your home?

Text-to-Text

Have you read other books about safety tips? How were those tips similar to or different from the tips described in this book?

Text-to-World

Firefighters help people in a fire emergency. What other kinds of people help in emergencies?

Glossary

accident – something that is not planned and often harmful.

detect – to find or notice something, especially something hidden.

emergency – when something unsafe happens and calls for quick action.

escape – to get out of a dangerous place or situation.

fire extinguisher – a tool people use to put out fires with a chemical foam.

Index

accidents, 10

candles, 6

escape plans, 14

fire extinguishers, 9

flames, 4, 18

lighters, 6

matches, 6

smoke alarms, 13

Online Resources

popbooksonline.com

Thanks for reading this Cody Koala book!

Scan this code* and others like it in this book, or visit the website below to make this book pop!

popbooksonline.com/fire-safety

*Scanning QR codes requires a web-enabled smart device with a QR code reader app and a camera.